Eduardo Argentino Campi

Temporal Laws of Nature of Pre-Columbian Mayan Science

Eduardo Argentino Campi

Temporal Laws of Nature of Pre-Columbian Mayan Science

The Scientific Modernity of the Pre-Columbian Maya

ScienciaScripts

Imprint

Cover image: www.ingimage.com

This book is a translation from the original published under ISBN 978-620-0-01606-5.

Publisher:
Sciencia Scripts
is a trademark of
Dodo Books Indian Ocean Ltd. and OmniScriptum S.R.L publishing group

120 High Road, East Finchley, London, N2 9ED, United Kingdom
Str. Armeneasca 28/1, office 1, Chisinau MD-2012, Republic of Moldova, Europe
Managing Directors: Ieva Konstantinova, Victoria Ursu
info@omniscriptum.com

Printed at: see last page
ISBN: 978-620-8-59847-1

Table of Contents

DEDICATION;

This work is dedicated to the Historian of Science J. Babini.

PROLOGUE.

The material presented in these texts was brought to me by its author, one day at the beginning of November. It should be clarified that at first I wondered (with some misgivings) if I could understand, without having previously studied the subject, what Eduardo appreciated and wanted to communicate about the Pre-Columbian Maya and the History of Universal Science; a question that was revealed to me with the voracious reading of these interesting pages.

Here, without a doubt, the interested reader will have the necessary basis to be able to understand the marvelous contribution made by the Pre-Columbian Maya to the Natural Sciences.

At the beginning, a comparison of Pre-Columbian Mayan Science and current Science is presented, highlighting the achievements of the former in the discovery of Mathematical Laws and their application to Agriculture. Then it is introduced in the prediction of phenomena, on the part of the Pre-Columbian Science, making allusion to the remarkable discovery of the Mayas on the Seriality of the eclipses and to the cultivation of several scientific branches, carrying out a great scientific task (not recognized), which were expressed by means of four remarkable objective laws.

The first of these Laws of Nature is that of the foliar implant of corn (Agriculture), discovered by the Mayas long before the Westerners, where it is compared to Galileo's Law of free fall of modern physics. It is here, where a similarity is observed in obtaining the representative numbers of its series. In this leyse becomes evident that they were some advanced in the matter and that la same constituted its first step towards la modernity. It cannot be denied then that, between pre-Columbian and western science, the similarity is the quality of modernity.

The second law of reference is that of the infinite powers (Arithmetic), where it is shown that the Maya reach mathematical or potential infinity through different eras: a fact more than evident to affirm that the pre-Columbians were formalizing the idea of eternity.

As for the third law called Integrated System, Eduardo evidences that Mayan astronomy was a Science and crowns the end of the chapter on the possibility that a partial heliocentrism was reached by means of a thinking of exact equations.

The last Law of Classical Calendar Science is that of Temporal Conservation, how did the Maya manage to estimate time accurately (conserve it)? The answer is in the reading of this great text, but today I could only advance that the prodigious Mayan calendarists already had a powerful modern calendar in ancient times.

To finish this exciting reading, the Universal Law of the First Digit that the Maya intellectuals did not explicitly state, but their numbers begin to fulfill it (applying science). In this it is said that when a number is greater the properties of the system are governed by simpler laws.

As for my initial fear, of not understanding the subject, today I could say that I am convinced, from these brief texts, that the contributions (not accepted) of the Mayas in Agriculture, Mathematics, Astronomy and Calendars were amazing from the pre-modern to the modern level; making it the modern western science that resembles the pre-Columbian one.

Undoubtedly, these writings will leave nothing to chance and this great writer, scholar and professional will not only be able to convince me but also prove to me with accuracy everything related to Pre-Columbian Mayan Science and its great contributions of modernity to Universal Science, as he will also do with you.

Concepción del Uruguay, November 11, 2024.

MarisaViviana Romero

Specialist in Science Education Professor in Mathematics Professor Researcher and teacher at UTN, FRCU Teacher at Universidad Autónoma de Entre Ríos, FCYT

ACKNOWLEDGMENTS.

Personally, I have had an appreciable career as an author, who has been able to publish a series of titles, in relation to science and its epistemology.

In recounting this story, it would basically be as follows.

(1987) "Epistemology of Clinical Psychology" (J. F. Kennedy Faculty) was published. We schematize up to the present the contribution of Freud to psychology. This is from Hypnosis to the psychoanalysis of human personality. Although psychoanalysis, as part of science, will have to improve some of its internal parts, its contribution to psychology is notorious.

(1988) "The Role of the Sociologist in Argentina" (Arts and Sciences). In this work we detail the constructive role of professional sociologists, with respect to their discipline. Which, in spite of BUNGE's criticism, has to be completed with the help of the rest of psychology, as VERON explained years ago.

(2010) "Pre-Columbian human rights to scientific creation" (secretary human rights of the Province of Entre Rios). Here we already defended the Pre-Columbian Right (as requested by MORGAN), regarding scientific creation. This is what we are publishing today and claiming for the Mayas. This was also recognized in the BUDAPEST Declaration (1999). If we take into account what MANCISIDOR (UN) explained, that only in 2013 the Science Magazine published an article of this type, we were not so far away in time.

(2019)The first study, "Psicología integrada de la Evolución de la Personalidad" (Editorial Académico Español). This study - to our knowledge - is the first to combine the complete (affectivity, intelligence) and systematic (baby, early and second childhood, adolescence) integration of the theories of FREUD and PIAGET (we personally belonged for a time to the SEPY: Society for the Integration of Psychotherapy). The Child-Adolescent Psychiatrist, gave us his work on the subject.

(2020) "Scientific Modernity of the Pre-Columbian Maya" (Editorial Académico Español). In this work we elaborate the Thesis of the scientific creation of the pre-Columbian Mayas. We began with the task of studying centrally, the different calendars (Lunar, Solar, Venus, Mars, Short and Long Count), and their resulting epistemology, elaborated by those people. We were beginning to have in clear, that the first scientific Modernity of the World, was carried out by people of Cacau skin (not Caucasians).

(2021) "Scientific Laws of Human Nature" (Editorial Académico Español). The author TAYLOR, referred that if Laws exist in a scientific field (Physicochemistry, etc.), we should also find them in other fields of study. Here we begin to study them in Evolutionary Psychology, Social Economics, History of Science and biology of species.

(2022) "Leyes del Modo de producción de los Conocimientos Científicos" (Spanish Academic Publishing House). A history of comparative science between East and West is developed here. That is to say, the most notorious epistemological characters of both orientations are highlighted.

At present we are in negotiations to publish (Editorial Académico Español), the brief work "The Laws of the Temporal Nature of the Pre-Columbian Mayan Science". This should be considered as the crowning of all the intellectual history presented so far.

PREVIOUS WORDS of the AUTHOR about the Mayan Society.

This is a small book, which is essentially about the History of Comparative Science. However, the only calculations you will find here are the basic arithmetic operations (addition, multiplication, etc.) with which the Maya operated. So, although this work is intended for specialists, with a little patience, it is easy to understand for ordinary people.

We recognize as important here -because of the coincidences-, the work of two Historians of Science; SANTILLANA-DECHAND: 'The Mill of HAMLET' (they call the first 'archaic scientists') and that of the anthropologists GRAEVER-WENGROW: 'The Dawn of Everything' (they recognize that in the villages the construction of the initial science began). In addition, for our own work, on the Maya, we are aware of some inaccuracies (that we have corrected here), which will always be minor in importance, before the omission of the scientific Modernity, achieved in the Classical Calendaric Science, by the pre-Columbian peoples.

The central thesis is not in this history, which scientific branch (physics or calendars), is superior or better, since this depends on the purpose for which one wants to use this science. Here comparatively we will verify in which case (oriental Mayan society or western northern hemisphere), an important quality, as modern science, was developed first, in its respective branch of studies, with written Laws.

The historical Mayas represent meritoriously all America, but especially Latin America (Argentina), which was the place, from where this new unpublished discovery (thesis on modern scientific laws) was enunciated.

Before going on to support our thesis, it is convenient to refer to some of the general characteristics that the historical Maya (as an oriental society in their way of life), had or did not have.

In the first place, they never constituted a unified Empire, because they only complied with partial alliances: TIKAL-NARANJO or KALAKMUL-CARACOL.

Nor did they live (if we follow architects like CHUECOGOITIA) in cities (civis or burghs), since the superior political unit was that of the Agrarian Ceremonial Center , which was governed by theocratic dynasties (COPAN: Sky; PALENQUE: Shield, etc.).

The Maya, through the QUICHE people, write the POPOL VUH, which has been considered similar to the Bible. There they forbid human sacrifices (GIRARD), which had to be replaced by serpent butterflies. In conclusion, to the two territories (classic Maya - present-day West), it will give them so much work (in spite of the commandment), to prevent the death of the people. Thus, if one region is civilized in these terms, the other will also be so; or, both are archaic in this sense (barbarism). If this moral equality is not taken into account, it is an uncorrected error, which even anthropologists make (Página 12 newspaper article), but they attenuate (in spite of our warning) to their true examples (it is the same denial to maintain that in the West, the reference of the commandment not to kill does not exist).

But one of the major shortcomings - in our opinion - that is committed with the pre-Columbian Mayas, is not to explain in detail, how they without developed instrumental technique (in this they were in the stone age), managed to reach the classical society and the establishment of their High Culture (which without another alternative explanation, should not have happened). One of the few authors that analyzes this deficiency, is SHARER in his work The Mayan Civilization.

The basic question here is not why the Mayas fell (which we will explain later), but when they reached their crisis. That is to say, what happened to them that in spite of their meritorious compensation (which has to do with our work), they could not overcome the situation.

It has also been mistakenly written (MITRE 1949), about the intellectual capacity, something that touches the Mayas:

"*To think that with the elements* and *in that environment, could incubate* and *expand a binomial like that of NEWTON, a mechanics like that of LAPLACE, an inventiveness like that of FULTON or EDISON, would be more than asking for pears to the elm. It would be like hoping that the characters of the printing press could be in the hands of savages,* and *coordinated by them, in millions* and *millions of ways, could be born the Divine Comedy of DANTE, since intelligence would not preside over the operation. That is why without the principle of fecund life that inoculated the blood of European civilization, the American man would have vegetated like his trees, without assimilating new reproductive forces, like the savage of MONTESQUIEU, who cut down the palm tree to gather the fruit. God did not give the American Indian the opportunities with which the superior races carve out their own destiny* and *enlarge the foundations of transcendent genius".*

As we will explain in this paper, almost all of this fragment is erroneous. This is the great superstition, which the West has elaborated about the indigent pre-Columbian savage (Indian), without correcting it mostly (this is in spite of the exceptions), until now. The intelligence coordinated very well the Mayan work, as well (this is with successes and errors), as in any other part of the world. This was largely due to the fact that they took vegetables (plants, trees) as an archetype. If all this were already accepted exactly, our present work would not have appeared.

The Maya, like almost all the peoples of the world, began intellectually, with a pre-modern stage, based on the '4/5 Elements'. This set of meanings is composed - as its name indicates - of two complementary parts united: the numbers (4/5, etc.) and the Elements of Nature (Water, air, earth, fire, wood).

These aspects were embodied in the glyph number 7 MANIK, of the Central Calendar on the Counting of Days, which was an image of a Hand. It represented at the same time the productive social work (Artists - Artisans) and the numerical base of the fingers to count (it is not by chance that the basic vigesimal system was derived from two pairs of joined hands).

The central point in this work will be the numbers (4, 5, etc.). The Mathematician STEWART referred that in history the number 4 (CAN), is the one that best represents the Elements. These numerals were applied to real Nature, hence the Mayan numbers (positive integers), come from the field of the Naturals (HENSEL). Thus it is not random that numbers, at present, are a component part of science (Arithmetic).

The Natural Elements (Air, water, earth, fire, wood), were related to social production, in which the village people, with their homemade tools (pots, mortars, etc.), began to build objective knowledge. Water (HA) was used for agriculture; fire (KAK) was used in cooking; the element earth or stone (CAB) was used (lime kilns) for construction; and so on (Air: IK; Wood: TE, etc.).

This technical activity of simple application (instruments before the first wind machines), not only maintained society, but also acquired an added or heuristic value for the development of knowledge (science). For the homemade activity that almost all people could perform, within the appreciable reach of their hands (practical work), was displaced in its causal scheme to Nature not directly perceptible (invisible) or out of reach of the hands (remoteness).

Thus, the celestial plane (night sky) was thought of as a river, along which the canoes sailed, carrying the stars (related to navigation techniques). The properties of the cooking fire were transferred to the sun (light, heat). The dispositions of stones and measures of the surveyors or builders, will give form to the spatial scheme of the universe (staggered floors). The domestic plants were the paradigm of human knowledge, since people grew from below (floor) upwards (the cob often represented the head of human beings).

Thus a relation of causal similarity (not of forces) was supposed between Nature and Society, and vice versa. Although this procedure is not always correct, we should not underestimate it, since several of its items reach the present; Corn, cotton, Mamon, Cigars, Cacau, Chicle, Rubber, some chemical and medicinal substances (this activity is recognized by UNESCO).

With the passing of the centuries, the manual part of knowledge was perfected and became known as: experimental (variable controls). It is not by chance that the word Elements appears as a title in several works of science (Geometry, Physics, Chemistry, etc.). Even common people sometimes still refer to the action of the elements of Nature.

When the Mayas reached their intellectual or scientific zenith (in the III-IV centuries A.D.), the different elements were included in a Mathematical-Arithmetic law, since all of them (Air, Water, Earth, Fire), represented in the summit of their monuments (stelae), equally fulfilled the numerical temporality (law IV). Thus in the upper part the celestial space or sky (Air) was symbolized by (5) CHICHAN (flying serpent); in the middle part the Water (represented by part of fish), was noted calendrically as (12) EB; below the Earth or stone was the (17) CAB and in the last part below, was the fire (solar circle), indicated in the calendar as (20) AHAU. All this together, began to count (counting to the Center) at the 5 Eras (see Law II).

The sense of being 'primordial' (raw material) is still maintained for the word 'elements' in the field of natural sciences.

"The Maya may become of interest to the Historian of Science" M. COE.

PRECOLOMBIAN MAYAN SCIENCE and PRESENT-day SCIENCE.

Of course, the science of the present is not the pre-Columbian Maya Science, but the reciprocal is also true, because in the past, the Maya Science obtained achievements, which were not found even in the West. In addition, some calendar calculations are still more accurate than those of the West.

Mayan science, with the same operational (formal-empirical) logic as that employed later in the West, passed through the main stages: pre-modern, modern and enunciation of Laws. When Western commentators have referred to an 'unscientific' stage, it is better to think of premodern, and when we are told 'marvelous' or 'interesting' the adequate qualitative term is 'modernity'.

The scientific historical process will take place continuously in a society with an Eastern way of life (GOLDEN), which oriented the mode of production of scientific knowledge towards another branch that was not Western physics (the Western social influence on physics was recognized even by HAWKING). Thus, in America the temporal concordance (synchrony) would be studied and in the West mainly the Force (HP).

Few authors realized (SPRAJT, AVENI, COE), that the Mayas thought to a science (Calendrical) and still less, none notices, the comparative height that was achieved. This intellectual activity was a Science, because it already had in itself the 3 corresponding planes of the present one. The Technical application (Agriculture), the control of variables (Luni-solar periods, etc.) and finally, the major Theoretical conception (infinite time).

The English word 'science' is not the first word in the world, to define the scientific meaning, so we will use here, some of the Mayan words, which are closer to these adjectives (in the future we will be able to find others more accurate). Thus the pre-Columbian Science (CHUN YAX) will not only discover Mathematical laws, as defined today by western scientific authors (CARNAP, PICKOVER, etc.), but also express them verbally.

MERTON for Modern science noted (we will take this definition as the basic comparative reference for the Maya Laws):

"It is necessary that the phenomena be conceived in such a way that the heterogeneous is reduced to homogeneous quantifiable processes. The repeated elements are indexes of a Law. Once the phenomena have been reduced to order, to a common unity, they become manipulable" (I).

As we will analyze the Mayas fulfilled almost all these requirements (TZOL: Order, enumeration, etc.), therefore, that definition of scientific Law, will be our model (MOCH), to apply comparatively in the Mayan work.

The application corresponds to Agriculture. This activity will be mathematized by means of accounts (operations), which in total lasted in Mesoamerica 260 days of the year (SPRAJT). For example, the Mayan Agricultural calendar (GIRARD) began on February 8 and ended on October 25 (February 20 + March 31 + April 30 + May 31 + June 30 + July 31 + September 31 + October 25 = 260 days). Depending on geography, these dates could be displaced, thus achieving at least 2 annual harvests; 1) 52 days (30 April-21 June) and 2) 73 days (13 August-25 October).

In spite of the fact that the Mayas did not have technical instruments (draught animals, wheel, metal plow, etc.), since they used a stick to plant with a fire-hardened point (XUL), they produced a productive revolution (Nobel prize economics LUCAS). This happened, because what they did not have in concrete Technique, they overcompensated it with their applied calendrical science, to Agriculture (as it is said today in the West: 'agrarian sciences').

It is necessary to think to explain this unknown, that while on the plants no force can be applied (JUILLIEN), the Europeans already had - thanks to the aeolian technique -, the horse steam (CV), and that 4000 CV, were equivalent to the work of 40.000 people (BRAUDEL noted 10 million Horse Steam, available for the West). This makes it impossible to try to equal this productive force, based on group work (it would require a non-existent overpopulation in the Mayan region, as it would rather be that of the entire American continent).

The classic Mayan civilization will collapse (war with prolonged chronic drought of 3 centuries, from the XI century of.), when this numerical chance calculable in its averages (rainy seasons), was transformed into 'wild chance' (MALDENBROT). That is to say, when the methodology referred to above (planting season, rains, harvests) could not be correctly followed or stably prevented. Nevertheless, from these historical operations, for maize plants, a (first) Law will emerge.

To continue with the plant theme, to the control of variables (experimental), the same DARWIN emphasized the importance, which is given to the 'breeders', including the domesticators of cultural plants (such as the Mayan Corn). With these crosses, it is a matter of achieving a cumulative effect (during successive generations), starting from almost invisible qualities. The English author refers that only 1 person in 1000 possesses this creative capacity, and adds;

"If endowed with these qualities (visual and reasoning), the person, during all the years of his life, studies with unwavering perseverance, will succeed in obtaining improvements" (2).

For the same reason, current specialists (FEDEROFF, etc.), refer that this was the first feat of genetic engineering. The geneticist DOBZHANSKI refers that, in some cases, the organic combinations (crossbreeding of specimens), are greater than the number of constituent particles of the Universe. Thus the Mayas selectively co-operated with the creative power of the evolution of species, achieving an increase in size, number of grains, etc. (the plant hormone Rubisco is one of the compounds found in the greatest quantity on our planet). MAGENDORF refers that they have been found in the Mayan geography (KUAUTEMALLAN), to all the dispositions of grains in the cob. BEERLING has explained that vegetables are the only ones that incorporate energy from the outside (sun) to the chain of Life. ROSES-CHUA add that "plants are machines", being this organic artifact (BENZ), composed by Humans (Corn), the one that helped them to produce on the scale already mentioned.

In most cases, the Laws of Science, in their scheme (KOHEN), are synthesized from the religious law, when their moral obligation (duty), is transformed into a numerical rule, which is fulfilled arithmetically (objectivity). Some Mayas achieve this, from their orthodox religion (BRODA), even if it is so wanted, with partial deism (because they were ComoCOPERNICO priests).

We have already explained that the pre-Columbians wrote down their laws with arithmetical formulations, which does not mean that they are not science, the proof is that these numbers in their results coincide with some initial formulas of later modern European science (physics, calendars).

In this work of history of science, we will continue the path marked by numerical operations, because they, like those of western science, depend on the previous existence of mental operations (thought) that produce them (there cannot be arithmetic operations without previous mental operations that have created them). Also for the Mayas these two operative planes were companions or went together (CUATES). The psychological equilibrium of the different mental meanings is given when 2 psychological operations (one direct and the other inverse): X + Y = Z; Z-Y = X (where the first meaning X -whichever for the subject who thinks it- remains equal to the final X; that is to say, X = X). Mathematician GAUSS explained something similar for the formation of the number zero (0), when he wrote + 1 -1 = O. The same in the version of Mayan operations would be:

(KITAN) + (HUN) 1 (KUNAH) - (HUN) 1 (LUB) = (MI) 0 .

Psychologically (PIAGET 3) the different mental operations are derived from the internalization of human behaviors, with different objects (+: to join, -: to separate, =: to exchange, etc.). These actions in Maya are +: YUC -and- -: TZUC (opposite and complementary movements). These mental operations by conserving the objects de-center the persons, of the error of the appearances (for example 5 stones in row, continue being equal in quantity, than 5 stones grouped).

In this way we will study the exact operations (LUB=), carried out by the pre-Columbians, because they will demonstrate the adapted mental decentration (objective), which they achieved scientifically (as any other scientist in the world), when elaborating the separation from the error, from their subjective I (asymmetric contradiction not compensated).

Thus the Maya not only placed numbers in their particular names (7 Serpent, 0 for the calendarist priests, etc.), but also humanized the numbers (faces of persons, etc.), . So their monuments of carved stones (stelae), with human figures surrounded by numerical faces, were a rustic literal x-ray of the human spirit, showing the significant internal operations of Persons.

The numbers were Alters Egos (anthropologists) or their other Psychological Egos

(SIITS). As PTOMKIN referred, in Mesoamerica a new (numerical) economy of the Human Mind is recreated, in which the anatomical differences (body parts such as faces) are mental (AUSTIN). For example, the number zero (MI) was indicated by a face taking the chin (pear). Thus the MAYAS were the only ones to have a numerical system with human heads (which in a way humanized arithmetic).

But the faces will not only maintain a profile perspective (flat), but also a dimension of carving in 3 dimensions was added to them (which is appreciated in their architecture). Thus in the construction of buildings (UXMAL, CHICHEN ITZA), the corners, showed faces (masks), with their real halves towards the back of the observer, who is appreciating them, from the vertex of the construction. This was just the opposite of the European Renaissance painting, in which the real plane of the picture (canvas) was forward, and the virtual depth, in perspective, was backward from the plane. So that if the observer stood at the side of the construction (body parallel to one of the walls), he only captured one of the halves of the face (partial face). To perceive the whole face, the observer had to stand, in the position facing the vertex (of the angle of the two walls), with his body parallel to the tangent passing through the angle of the corner (if he did this, the mascaron also looked at him).

Of course we do not affirm here, that the Classical Calendar Science is superior to Physics, nor vice versa, since there is not a scientific branch greater than the other (since all serve), depending its value more, for what is required to the respective scientific knowledge (time, force, etc.).

Then, following an initial example, we will give 4 more in total (one for each branch of study), which will be sufficient to demonstrate what has already been referred to. In other words, that it was constituted with the statements of Laws, a Modern pre-Columbian science. This quality fits for this, because most of the Mayan arguments are associated to later modern examples of Western Science.

INTRODUCTION; THE PREDICTION of PHENOMENA.

This is a writing about pre-Columbian science and the laws it obtained. Although there were contributions from other peoples (OLMECAS, ZAPOTECAS), it is specifically about the Mayas, who obviously embodied a broad society (writing, religion, etc.), which cannot be denied (perhaps the least developed of it was the technique). The inverse (as several authors do), to omit denying that they also practiced a scientific task, is a great error, close to discriminative ignorance. As it is basically about Arithmetic applied in different fields, its simplicity, makes null the excuse of incomprehension of the progress obtained, if we refer to the History of Comparative Science.

For some Maya intellectuals, who, thanks to their mental operations, moved away from religious reductionism, not only cultivated several scientific branches (applications, arithmetic, astronomy, calendars), but also expressed this knowledge by means of objective laws. This is what we will basically demonstrate.

To give an example of modern quality, we will quote POPPER (Logic of scientific research), when he explained; "the task of Natural Science is to look for Laws that allow to deduce predictions"(4). This will be what the Mayas will do in their field of studies (prediction of seasons, etc.), even in an exact (numerical) way.

In the branch of their astronomy, they not only obtained a science (AVENI), but also fulfilled those advanced characteristics, that is, the ability to predict events scientifically, before they happened (on the other hand, in the notion of orbits, they were just as wrong, as was COPERNICUS himself).

In this sense, it was then an 'expert' astronomy, as it is recognized today in the West, for astronomers to be able to fulfill the task of predicting eclipses. At present some astronomers are recognizing that the historical classical Maya were perhaps the only ones in the world to be able to foresee these astronomical events.

In order to be able to do the same, you have to have, with exact data of Solar and Lunar periods. The Mayas in their Uinal glyph joined the sun with the moon. The chronicles refer that it will be 'night and day at the same time' (solar eclipse). The Mayan astronomers will start from 405 total lunations or 33 solar years (it is probable that they have observed the sunspots, through jade sheets, because this number is a multiple of their period: 11. 3 = 33).

Thus we have the following equation, for the tables used for the eclipses: (table)

(Sinódico lunar) 29,53 días. 405 = 11959, 65 = (semejante) a 11960 días (en Maya 1.13.4.0).
Entonces anotaran a 3 grupos distintos de cómputos Lunares:

9 (c/u: con 3 de 30 días; y 2 de 29).	148 días (subtotal).	1332 días (total).
53.(c/u: con 3 de 30 días; y 3 de 29).	177 días.	9381 días.
7 (4 de 30 días; y 2 de 29).	178 días.	1246 días.
------ .	.	-----------------
69.	.	11959.

At the end of each group (69), an eclipse may occur (GOLDEN). These tables were correctively adjusted every 50 years (THOMPSON). The central discovery will be number 173, 3 days (TEEPLE). In COPAN the stele M was engraved with this number: 520 / 3 = 173. For us this number derived from 11960 / 69 = 173, 3 days. This number is the temporal translation, of the later geometric nodes, of Western astronomy.

Of the 5 intervals (days), that today western astronomy has, for solar eclipses, 3 of them were already recognized by the Mayas: 1039, 1211 and 1742 (RAMOS). Thus in the previous table, which began on the day 12 Lamat, the following operations were fulfilled: 12 Lamat (168) + 173 = (341) 3 Imix; 3 Imix + 173 = (514) 7 IX; 7 Ix + 173 = etc. For the date on which the eclipse fell, we had a margin (standard deviation) of 3 days. If the date was closer to the full moon, the eclipse was lunar, if it was closer to the new moon, the eclipse was solar. They also knew that an eclipse could occur in another place, even without being observed from the Mayan geography.

They also knew (COPAN 837 of. stela A: 18.5.5) to a period of 18 years ((6585 / 365,242 = 18), to which 10 days were added (6585 + 10 = 6595), in which, the order of all (70) eclipses is reiterated, again (29 lunar and 41 solar). In the middle of the XI century of. was erected the Temple XI of COPAN, in honor, to the discovery of this serial order in the eclipses.

We will now proceed to the statement of the 4 Laws, previously referred to.

CHAPTER 1: LAW I: FOLIAR IMPLANTATION

(Agriculture: U CHUL WAKAL KIL)

The Mayas did not know the laws of genetics (MENDEL), but instead they initially discovered the Law of the phenotypic disjunction implant (Maize), of the appearance of leaves (BIL in Mayan) long before it was completed by Westerners (FIBONACCI, BRAUN and other physicists (5)). Therefore, we are already in the presence of a law of Nature, it only remains for us to find out the level of advancement it had (as we will demonstrate below, because of its advanced operative structure, it is a Modern Law).

The law (operations ordered in series), here was given by a numerical succession of arithmetical operations, which registered a series of causal events (I°, 2°, 3°, etc.), phenotypical of the growth (days), of Corn (NALTEL). The explicit numbers (days) were as follows:

1) 4
2) 5
3) 8
4) 13
5) 20
(....)

As these numbers (series 4,5,8,13....) are not found in the international registry of Integers, there is no doubt that they were created by the Mayans.
Here the phenotypic changes observed by the Mayas (GIRARD) are: I or) at 4 days after planting the grain (NAL), la seedling (HURAKAN), la despuntaba; 2 or) at 5 days sprouted la first lateral leaflet (symmetry break); 3 or) at 8 days, comes out la 2nd. 4°) at 13 days the corn (IX IM), has both leaves approximately the same length ('parrot wings' as the Mayas used to say); 5° at 20 days the vigesimal unit (UINAL) was reached, etc.

If now (we remember that Americans did not practice physics), we match biunivocally (1 to 1), these numbers with those of the formulation (E = Tal square (6)) of Galileo's law (one of the first of modern physics), of free fall, where the space traveled by the body (1,4,9,16...), is equal to the square of the elapsed time

(0,l,2,3,3,4...). Thus, the following comparative table is obtained:

4	0
5	1
8	4
13	9
20	16
(.......)	(......)

One has that implicitly both series of obvious numbers (implicit, fall), are produced by implicit additive operations (+) with equal numbers of the series (1,3,5,7,...) of the odd ones (4+1= 5 and 0+ 1=1 ; 5+ 3 = 8 and l + 3 = 4;8 + 5 = 13and 4 + 5 = 9; etc.). This is called isomorphism in Arithmetic (literally equal form).

This temporal comparison is allowed, because the Mayas privileged the odd numbers, thus operating without knowing it consciously, with prime numbers (3,5,7,11,13,....), because the only one of them that is even, is the number 2 , which also appears forming a of the implicit common odd numbers (1+2= 3; 3+ 2= 5, etc.). They could also draw the least common multiple. Thus it is not by chance that it has been possible (CARLSON), to factorize the main Mayan (calendrical) numbers. Although the numbers in the given example, which produce the others are dimensionless (BUKHIMGAM), both series are regulated by the elapsed time (T).

If we take into account the chemical periodic table -another law of Nature (here the molecular symmetry is called isotactic)-, we notice that it is also constructed in an orderly fashion (although each element is causally different from the other), by means of the implicit interaction +1 (1+ 1 = 2; 2+ 1= 3; etc.), unlike the series of odd numbers (primes), already detailed.

In short, the Mayas were experts in finding and highlighting the integers (a difficult matter to achieve) in Nature: snake of light at equinoxes (day = night); 5 and 65 Venusian synodic revolutions (2920; 37960); in COPAN some buildings were turned 7 degrees 40 minutes north (AVENI), which suggests that they were guided by the magnetic declination (before the Europeans knew about the magnetic inclination), and finally, although they did not know the number that regulated (FOURIER equation), the phenomenon of echo, they did reproduce it masterfully, as the song of the QUETZALT (LUDMAN).

Concluding here the Law (1) Mayan operative arithmetic of the implant, it will be essentially noted as "13 + 7 = 20", because the Mayas had the habit of simplifying. Or so they enunciated it:

(13) OXLAHUM (+) KITAM (7) VUC (=) LUB (20) UINAL.

This Law was related to the calendar of Agriculture, because 365 - 105 (30 April to 13 August) = 260 = 13. 20. The numbers 13 and 20 were explicit numbers (of the observable growth of the Corn) and the odd number 7, was implicit. Thus 7 + 13 = 20, implied the whole series of numbers already exposed.

It is not by chance that the Mayas called the number 13, as KUL or computer principle (HASSELKUS). Thus on the pre-Columbian side one had Arithmetic and Biology (vegetable), and on the western side Arithmetic and Physics. For another similar case, it has been referred (GOODWIN: 7), that Mathematics, Physics and Biology were combined.

Classical physicists (since CHURCH) have carried out more complete experiments (plants), but the Americans were centuries ahead of them in initiating these studies. Of course, in none of these two geographies, there was any knowledge (except GALILEO, who intuitively knew something), of what was achieved in the other (let's think that the West still does not know much about the Mayan contributions).

STRAUSS spoke for the pre-Columbians of a 'vegetal Pythagoreanism' and the Nobel Prize winner WILCZEK (8), also added to the case of a 'Modern (arithmetic) Pythagoreanism' (it is not superfluous to say here, that he also referred that Pythagoras enunciates one of the first laws of acoustics; we agree that, in his case, it is that of a pre-modern law). However, for this Maya case, they were already effectively in a Modern Law, because as mathematicians refer (NAVARRO, TREJO: 9), the operative isomorphisms (1,2,3,5, etc.), are to establish integrated theories. In this example we emphasize that what is common to both scientific branches is the quality of modernity. It can be concluded then that the quality of advancement of both series (modernity) is true.

For if one looks at the series (isomorphic numbers) from one side or the other, one notices that there is a symmetry (from both the same quantities are contemplated), or as it has been explained in Western science, that the fundamental Laws are symmetrical. In other words, Nature (physical or vegetable), for the basic mechanisms, uses symmetry (WILCZEK), that is, invariant (equal) quantities. They represent 'any' perspective of the observer. Thus the quantities are conserved (NOHETER) in principle (in psychology this isomorphism is called or is produced by reversibility).

This is totally coincident then, with the statements about agriculture (plants), when it is affirmed that it was a science for the pre-Columbians (MANTEGAZZA) or that it was the first step towards Modernity (WHITEHEAD).

CHAPTER 2: LAW II: POTENTIAL INFINITY

(Arithmetic: WOOH)

At this point we will develop the theme of operations coordinated associatively among themselves. In arithmetic *'1* + 13 = 20', it was expressed as: '7 (VUC) + 13 (OXLAUN) = '1' (HUN) '; or 7 + 13 = 20 UINAL or vigesimal unit. The Mayas will make before so GAUSS refers it, to la Arithmetic queen of the sciences (because they apply them in all their branches). These operations formally approached the sense of law in the current mathematics, when one speaks of operations, as Laws (commutative, associative, etc.). It is not by chance that WHITEHEAD consequently explained that arithmetic is a feature of modern thought.

It is for this reason that similar elementary formulations (1 + 1 = 2), were included by the mathematician PICKOVER, in his work on 'Laws of Science', within a brief list, for their capacity to have transformed the world. We will develop then, as the vigesimal operation, will make the pre-Columbian world immeasurable, through the operative interaction (MERTON 1). The vigesimal system had a major operative advantage over the western decimal, since in the 5th rank or step, there was already a difference of 134,000 units (= 144000 - 10,000).

As the Americans deploy their vigesimal numerical system (20 in 20), this quantity and its multiples (THOMPSON), will appear in almost all fields (agriculture, astronomy, calendars). It will start from small (MAR) quantities, arriving at large (CHAN), and from there to infinity (HUNAL).

For example, numbers (thirtieths or twentieths), and names (days, months), were at first separated (TZUK), and then associated (YUC).

Thus (by biunivocal correspondence) the base calendars will be assembled, the 13nas and 20 names will give the Tzolkin (260 = 13. 20) and 20 ñas with 19 names will give the Haab (365 = 18. 20+5). After 73 revolutions of the Tzolkin and 52 of the Haab, the quantities were equal again (73. 260 = 52. 365). Being born another calendar denominated WAKIL KILIL or Short Count (18980 units in Maya 0.2.12.13.0) and finally from a common date 4 AHAU (160 Tzolkin) 8 KUMHU (348 Haab), would begin the Maya masterpiece XOX IT or Long Count (which allowed them to have a Date Era from which, the Calendar Wheels would escape to the infinite potential). They would be added to count here large units Kines (1),

Uinales, (20), Tunes (360), Katunes (7200) and Baktunes (144000).

If we make an evolutionary scheme of all this development, we would have:

Numbers (13nas)	**Names(20)**	**Numbers(20nas)**	**Names(19)**
11	IX	2	KUMHU
12	MEN	3	KUMHU
13	IBC	4	KUMHU
1 CABAN		5 KUMHU	
2 ETZNAB		6 KUMHU	
3 CAUAC 7 KUMHU			
4AHAU 8KUMHU			
5 CIMI 9 KUMHU			
(....................................)			

The computations in the Long Count were subject to formal rules of their own. For example, every time a name different from the Month (HAAB) appeared, the same name of the Tzolkin reappeared; zero in the place of the Kines, the name Ahau reappeared, this also happened with the numbers (52, 73), which reappeared again, etc.

The Maya did not know that the above will be a synthesis of the mathematical 'Mother' structures (BOURBAKI), both of order (Short Count) and partial algebra (Long Count). But in practice they approached them. The balance in the first is reciprocity (from a first computation 11IX to any other 13 CAB, there is the same (=) quantity 2nd. that vice versa and in the second case, they are inverse operations (+1 -1 = 0). When psychologically (PIAGET: 10), the mental structures (concrete-formal) analogous to these (4 AHAU 8 KUMHU), are combined among themselves, that is to say that they operate together or simultaneously, possibilities of new qualities will appear, as when the real (empirical), is subordinated to the hypothetical (Operationsformales).

This was what allowed the Maya to reach the mathematical or potential infinity (BOLON TZAKABIL), through different Eras. This was done without losing accuracy, for example, each end of BAKTUN fell alternately on an equinox and solstice (HERNANDEZ); the 5 Eras were close to the precession of the equinoxes. From la archaeoastronomy has been contributed in this regard, since have been given evidence that almost prehistoric cultures (SANTILLANA), began to study this celestial phenomenon. This period is very difficult to calculate in its totality (23000 - 26000 years), having to deduce it from partial observations (MENZEL) (human life), with sights (ETZ), that pointed towards the distant horizon (as the Mayas did).

Making an average calculation, according to modern western astronomy of 25729 years, and extrapolating it for the Mayan epoch 25950 years, we will have 25840 years, which, with respect to the formal value of the 5 Eras of 25626 years, gives a difference (without correcting it: CASTELLANOS), of only 214 years (or u∩8 % of error to the total value). Thus the author SEVERIN explained that the Mayan Calculations of the precession of the equinoxes are found in the Paris Codex (MAUPONE).

Thus, if we make a schematic picture of the operations towards potential infinity, we will have:

1 Era **(WINAQ MAY KIN) = 5125 years or 13 Baktuns (13.0.0.0.0.0.), recorded in QUIRIGUA.**
5 Eras (HOKOL MAY KIN)= 25,626 years or 65 Baktuns (3.5.0.0.0.0.0): COPAN.
Great Era (CHAK KIN OCHTE)= 374,152 years or 949 Baktunes (2.7.9.O.O.O.O.O.O.) TIKAL-UAXAXTUN.

From 1366560 (9.9.16.0.) a division / 73 was made multiplying the result by 100, and the value of an Era or 7200 (Katun) was obtained. 260 = *1* Era. In days *1* Era; 1.872.000 . 5 = 9,360,000 days of the 5 Eras; and days of an Era 1,872,000 . 73 = 136,656,000 days of the Great Era.

For this reason it has been said that the Mayans stamped (SCHELE) to one of the largest finite numbers and added (DORADO), that the pre-Columbians were formalizing the idea of eternity, crossing (COBA), to all temporal limits. Let us think that comparatively to the Greeks it was difficult to reach infinity.

We find ourselves in agreement with the historians of science (KOYRE, GOLDSTEIN, etc.), when they express that from COPERNICO the universe became (Renaissance), infinite or spatially Modern; but it must be taken into account, that some Mayan thinkers, intellectually raised before the same thing, in the temporal parameter. It is possible that in some calculations they have surpassed the estimations of the current physics on the age of the Universe.

This helps to understand (BROTHERSTON), that it is the pre-Columbians who explain to the westerners, that the world had more than 6000 years of duration, idea which arrived until the European Enlightenment. Thus, although the Mayas received previous knowledge, they were the ones who took it the furthest (infinity).

CHAPTER 3: LAW III: INTEGRATED SYSTEM

(Astronomy: WAY HA)

In this case it is a question of co-univocal operations that are integrated (TZAY), in a system of total set. We already know (AVENI: 11), that Mayan astronomy was a science. This astronomical accuracy (which reached 1 day of error in 6000 total years), was achieved by observations, which escaped to the horizon (for example, the sun on the horizon had its own glyph).

In this branch would be noted a formulation, which was called (by the previous author), as the 'super number'; in Maya 9.9.16.0.0 (= 1366560 days). This quantity would articulate several celestial cycles (CUC) to each other.

The Law (1) of the temporal nature was stated: BALUM (9) BAKTUNES (= 1296000), BALUM (9) KATUNES (=64800), WAKLAHUN (16) TUNES (=5760), MI (0) UINALES and MI (0) KINES. Thus: 1296000 (= 144000.9) + 64800 (= 7200.9) + 5760 (= 360.16) + 0 (= 20.0) + 0 (= 1.0) = 1366560 .

In numbers of calendrical units (KINES = days), this large number above, coordinated to several equal periods:

-(Agriculture): 260.5256 = (Tropic Year): 365.3744 = (synodic Mars) 780.1752 = (synodic lunar) 29.53 . ⅛⅛TΠ = (synodic Mercury) 117.11680 = (synodic Venus) 584 . 2340 = (LUB) 1366560.

The Modern physical conception of NEWTON (XVIII century) will be the one that integrates (uni-verse) in the West to the celestial-terrestrial plane, by means of the gravitational force, which affected both the apple and the Moon. The Mayas also manage to synchronize through the absolute T (time) (t= t'), the sun (space), with the terrestrial corn (which is one of the most solar plants of the earth). We are demonstrating something similar (universality) for the Mayan case, because the Galilean law of fall, approaches in the terrestrial soil, the causal effect of the universal law of Newtonian Gravity (inverse of the square). Gravity is the common factor in both cases (corn, stone), although in one it goes upward in growth (the albumins capture this force) and in the other downward (descent).

So perhaps, as PRIGOGINE suggested, it is true that, in the Universe, life is as common as the fall of a stone (today it is known that plant quinones, which come from outer space, are similar to those found on Iatierra).

The historian of science KOYRE explained that the English author brought accuracy (unification) from heaven to earth (physical), and in turn, the Maya already had it not only in heaven (9.9.16.0.0.0), but also on earth (CAB). This is not only seen in the stone Stelae (4/5 elements), but in the formulation of their Date Era (2906 ac): '4 Ahau 8 Kumhu' (4 and 8 are Corn numbers and Ahau represents the Sun; SCHELE). In other words, Nature (symbolically in its legal universality), was represented not only in the Apple, but earlier also in the Corn. Thus in spite of KANT, in the figurative sense, the Maya are the 'Newtons of the leaves of grasses' or gramineae.

If we add that these numerical series (4,5.8, 13,...) of the growth of the Corn, are also produced by the Sun (photosynthesis), it is not casual, that if the same ones are continued under equal implicit operations (+ odd series), we arrive explicitly again to the main solar numbers (365, 260 = 365 - 105 days, of the 2 between zeniths of the sun). Schematically we will have, starting from the basic numbers of the Corn, a certain congruence, by prolongation of the initial implicit series:

4	+ 1=
5	+ 3=
8	+ 5 =
(.......)	(......)
260	+ 33 =
293	+ 35 =
328	+ 37 =
365	

If we also make an internal analysis of the quantity 9.9.16.0.0, we can find new truths, such as partial planetary heliocentrism, which includes terrestrial heliocentrism. The astronomer GUTIERREZ has wondered if the Mayas knew this very thing. As the anthropologist CASTREL asked to link COPERNICO with the 'savages', we will show here how the Mayas may have arrived at the notion of terrestrial heliocentrism, on condition that we keep in mind that the Polish author did not give empirical proofs of terrestrial translation around the Sun, the concrete proofs of this were all later (the author KOESTLER specifies that on average COPERNICO made to develop his meritorious work, 1 observation per year).

If the Modernity of the European author (which we respect) continues to be sustained in these previous terms, we will have the following explanation of the terrestrial heliocentrism, thought or elaborated by some pre-Columbian Mayas.

The Polish author said that, if human vision could be magnified, the planetary faces would be appreciated - in his time. This would be precisely what Galileo did with his telescope, when he focused on the planet Venus, concluding that 'the Goddess of love (Venus) has faces like Selene (Moon)'.

Something similar was also done by the Mayans, through the brightness (the pre-Columbians could locate it, even during the day: STRAUSS), of Venus (NOH EK). Some Mayan astronomers knew that Venus passed behind the Sun (superior conjunction) or emerged from the underworld (TEDLOK). This thought is stamped in the following formula (cycles in days), which the Mayas knew, because they really thought it:

(Venus) 584.2340 = 3744.365 (Sun).

The next step was to transfer to the Earth (CAB), this quality of Venus (partial heliocentrism). The formula that led to this was:

(Venus) 584.2340 = 5265.260 (Earth).

Thus, some Mayas knew before HERSCHEL, that although the terrestrial movement was not appreciable (Galilean relativity principle), since the earth moved through space (following the sun: PILEN), as its own name referred to it, in CAB the particle AB indicated the property of 'flight' (movement).

It could be concluded, more surely, that the Earth had a displacement movement or 4 Movement (= 260 / 65) around the sun, because CUC meant cycle (the same was for COPERNICO revolutionibus). This similarity with the solar motion (4 stages), was thought through the following operative formula:

(Earth) 260.5265 = 3744.365 (Sun).

Which is equal (through temporal kinematics) to what the astronomer Ptolemy once said, that it is the same to think that what moves is the earth with respect to the sun. But even more, because as the Nobel Prize winner WILCZEK refers, the numbers are for "any perspective of observer", including a terrestrial one that moves around the Sun, (the Mayas were decentered from the Earth system, with a spatial perspective, before LEONARDO does it). More specifically A. WILSON (11) has explained that as the Mayas knew the sidereal year, this implied an "observational mental perspective: from the Sun, appreciating the terrestrial movement". Which is equivalent to terrestrial heliocentrism.

In other words (MENZIES (12)), it is fulfilled here, what is expressed by the Encyclopedia Britannica, that there were some authors who also thought something modern astronomical as COPERNICO. For the case of Mayan astronomy, there is a possibility that a partial heliocentrism (Venus, Earth) was reached, by means of a thought (mental operations), of exact arithmetical equations.

CHAPTER 4: LAW IV: TEMPORARY CONSERVATION

(Classical Calendar Science: CHUN YAX OCHTE)

The Classic Calendar Science of the Mayas had several specific chapters (Moon, Sun, Venus, Mars, etc.) and at the same time linked to each other (Law III). The initial estimations are linked to the technique (METATE), but later their observations will become more specific (horizon, height), more in the sense of a Chronometry. Thus the calendarists measure very precisely the variable time (chronology), for example, the Temple of the Sun in PALENQUE, was literally an exact sundial (equinoxes, solstices, zenith). For example, for the Solar calendar (tropical year), the differences (=/=) to the reality by mistake, will be reduced with history:

VII -VI centuries BC: .... 0.24232 day.
Century II ac: ....0,00768 d (HUEHUETLAPALLAN). This was the correction equivalent to the later Julian of the West. Before Centuries
VI-VII of: 0.00028 day (ESTELA A COPAN). This was the equivalent correction to the later Gregorian of the West.
In the present: ....: error by one unit, it would not be in the fifth place after of the comma. The annual value (year of the tropic), would have a calendar value of 365,242 . 19 = 6940 = 7200 - 260 .

In the third place (before the VI-VII century AD), the Maya had already surpassed the later level of accuracy of the Western Gregorian calendar (Renaissance: 1582). This Calendar is the current one for the West and its directive composer CLAVIUS was called 'Modern' (a crater of the Moon bears his name).

But as this western Calendar, failed in 1 whole day in 3000 years, and the Mayan one fell in the same error, but in 5000 years, then we can specify also, to the previous Modernity of the American Calendar.

MORLEY concludes that the Mayans performed a prodigy in any Calendar system, either Ancient or Modern, of course if this happens, it is because already before, their calendar was Modern. In the amount of the Great Era (after 300.000 years, the supposed error for the location of 1 certain day, almost reached the experimental error of 10 places after the comma, in the most sophisticated scientific branch of the present West, : la quantum mechanics (SOKAL). This accuracy was obtained on the basis of the positional arithmetic system and the number zero (the West differently receives these achievements from outside its geography).

We remember the above, because to measure well, before (BACHELARD) it is necessary to conserve (KALAK KIN), the variable to be measured. Therefore, if the Mayan calendars estimated time accurately, it was because they conserved it: for example Venus 584 . 5 = 2920 = 365 . 8 Sun, etc. The Americans transcend with this, to the erroneous notion of surpassing a body to another (PIAGET), with independence of the trajectory traveled, concluding the equality (LUB) through time.

In this way, the pre-Columbians managed to preserve, in a permanent way, the duration of the natural world (GIRARD).

The concept of conservation (equality that is maintained in the face of apparent changes) arose (PIAGET: 13) long before any estimation of the laboratories of the Western northern hemisphere, which makes it possible that they have been discovered in other scientific branches such as the one referred to by us here. Spontaneously by the corresponding mode of production, the Maya children would arrive on average earlier, than Western children, in the conservation of matter (pottery) and also in the conservation of Semicontinuous quantities (corn kernels).

The Maya adults named the North Star as XAMAN EK, which was the guiding star for the traders who carried a bundle of cargo (KUCH) on their backs. This weight also symbolized the temporary load (AH KUCH), which was transmitted from calendar to calendar (stage to stage), to another relay (CHASQUI), which took it back carrying it (the previous amount was conserved). In other words, GALILEO discovers the conservation of energy (kinetic-potential) and the Mayas the conservation of time (past-future).

The Mayan scholars (classical calendarists), arrive at temporal conservation or the notion of absolute time, before the Westerners. This would also happen a posteriori with Physics (NEWTON's absolute time).

Today these invariant qualities are recognized by current science, for example, the molecular biologist MONOD explains that scientifically: "the most fundamental propositions are Universal Conservation postulates (14)".

Concluding what is a notion of Modern science in one side, it is also in another geography.

UNIVERSAL SET of MAYAN SCIENTIFIC TEMPORARY SCIENTIFIC LAWS

In this point -unlike the previous ones- we will detail the law that the Mayan intellectuals fulfilled, without knowing it or explicitly stating it (Universal Law of the first Digit: BENFORD).

From this universal law it has been explained (TAO :15), that when a set of numbers is larger (large), the properties of this system are governed by simpler laws. This is indeed what is true for the estimates of the Maya. In other words, the multiple complexity can be reduced to a few numbers of arithmetic, which is the field in which the Americans were operating.

Thus, although the difference to the theoretical proportional number for our few examples (N= 30), in this small work, is high 1.4 (= 2.9 - 0.9), because the normal deviation for BENFORD's law, is 3, but per 100 (percent): it must be taken into account, that this index improves, if the cases of the sample (large numbers) are increased. We in this work are in la sample (N) minimally to 70 numbers of this (by lOOto). However, if the numbers referred here were given by chance, almost all the ranges of the first digits (in the Mayan estimations) would have among themselves, a percentage of numbers quite similar (minor differences). For example 30/9 = 3.3 approximately in each one (for a total of 28.7), but also in reality (33.3; 13.3; 10; 0; 3.3; 3.3; 3.3 ;3.3; 0; 3.3) the average deviation is much higher: 7.7, that is, more than double the randomness. Moreover, as BENFORD's Law predicts, almost 50 percent (47.6) of the quantities would have been established for the digits 12, and the other 50 percent (52.4) would have been distributed among the digits 3 to 9. Our empirical results (30: 16-14) are close to this ***prediction.***

(1-2) 53.3, which deviates in 5.7 to the theoretical value, and 46.7 with respect to chance. (39) 46.6 deviates in average to the theoretical number 5.8, and in 23.1 to the expected by chance. That is to say that these counts of the pre-Columbian Maya measurements (noted in this work), are far from chance (they are for a reason that they are following). In other words, this universal arithmetic law, begins to be fulfilled, for the data of the Mayan Calendar Science, certifying as we explained, that they are part of a temporal Law of Nature.

What matters is that these Mayan numbers acquire (by complying with this legal norm), a truthful character, that is to say that they have few possibilities of being basically erroneous (since the numbers that are chosen artificially, do not adjust to this numerical law). In other words, they are false data and not taken according to scientific criteria. It is not superfluous to add that this Law (which we reiterate the Mayas did not discover), applies (as we do here) also (STEWART, LUQUE, TAO, etc.) in several scientific examples such as: geometric sequences, extended numerical successions (intervals of prime numbers), physical constants, a good part of the macroscopic physical laws (thermodynamics, fluid dynamics) and resonances of atomic numbers. Obviously the Mayas did not practice physics, but they did practice science, as we prove here.

CONCLUSION

The Pre-Columbian Maya and the History of Universal Science.

In spite of the fact that the Mayas inspire painters (RIVERA), architects (WRIGHT), sculptors (MOORE), and will give two Nobel prizes (ASTURIAS, MENCHU), there is something more in their contribution. Nor in spite of la importance we, we will not focus, in la productive agricultural revolution (GODELIER), almost to scale 'industrial' that I approached them (LUCAS nobel prize), to the economic product of the West, before la industrial revolution.

The historian BAYLI called this 'the great domestication' and DARWIN concluded the importance for civilization of cultural plants. Maize (IX IM) crossed (variety NAL TEL) by the Mayas (first feat of genetic engineering: FEDEROFF), would end up giving a female Nobel Prize (MC KINTOCH).

With all due respect to the English-speaking authors who detailed the religious essence of the Maya, we still think of another more relevant aspect. For some Maya intellectuals it was enough -if necessary- with a partial deism, to do what they achieved.

The revolution of the Mayas as it was referred some time ago (GIRARD), is about the history of the culture or as MARTINEZ wrote more currently (Geometría Mesoamericana FCE 2000), that: "it was a great loss for Humanity, the non-acceptance of their scientific thought. We do not have to subordinate to the intellectual level (ROJAS), nor let them steal our History (GOODY). When the first woman physicist SHEINBAUM assumes the presidency of Mexico, this great historical delay could be changed, otherwise we South Americans will be alone, defending this thesis (first scientific Modernity), against the ideological reductionism of almost all the Western northern hemisphere.

The objective Laws that we explain here, not only demonstrate that some Mayas worked in science (Agriculture, Mathematics, Astronomy, Calendars), but that all these developments, were linked at some point, with the Modernity, of the History of the universal Science. The Mayas crossed unprecedentedly and completely, in their branch of studies, from the premodern level, towards the Modern.

Thus we have to invert the comparative references (DORADO), because it is the modern western science, which resembles in something to the pre-Columbian one, and not the other way around. Otherwise, we will remain fixed in the old ideological reductionist Alliance (superstition), as explained by the Nobel Prize winner MONOD. The same author refers that society wants Science to serve it, but does not respect it (replacing knowledge by ideology is the greatest lie of all possible Humanities).

Literally the Mayan intellectuals were:

<u>The Original (first), New (modern) World (nature).</u>

Which refutes that, in the current Latin American geography, it is scientifically in the Third World (legend stamped in the Institute of Physics Trieste-Italy, in the attached institute of the History of Science for the Third World). In this sense we have complied with Professor BRODA's request, that is, linking what has been done in Mesoamerica with the history of general (international) science.

The conclusion of SEDEÑO is correct, when he referred, something that also applies to the Mayas: "the calendars constituted a corpus, so organized, as to be themselves, the theoretical framework, or a kind of paradigm or realization, recognized universal scientific'.

Today the history of science is correcting some mistakes that have been made. The historians SANTILLANA- DECHEND (Hamlet's Mill 6th floor 2015), called 'archaic scientists', the anonymous people, who begin to compile, one of the oldest branches of science of all: 'archaeoastronomy'. The anthropologists GRAEVER-WENDROW (The Dawn of all Ariel 2022), refute Marx, when they write that science began to develop in the Villages. Thus, as a group (as referred to in article 271 of the Declaration of Human Rights!), the objective properties of the Material Elements (weight, hardness, combinations, impermeability, buoyancy, etc.) began to be discovered. We correct NEEDHAM when he notes that a more commercial society (civis or western bourgeoisie), will be the one that, for the first time, reaches modern scientific thought (astronomy, physics), since the Renaissance (XV-XVI century), because the pre-Columbian , do the same (Classical Calendar Science), much earlier (at least 1000 years before). Not only they manage to elaborate cognitive values similar to the western ones, but they do it previously.

However, there are even Western Nobel laureates who pretend to overlook these merits. For example, WEINBERG in his work on Modern Science (explanation of the World), is mostly wrong, when he wrote that what was done in pre-Columbian America, was 'interesting', because its thought in advancement, was Modern. The author ELIADE explains that the West (first Europe, then almost all of the Western northern hemisphere), in developing its scientific path, believed that it was not only the best, but the only possible one. But he concludes that the magnitude of exotic values (foreign to the Western way of life), is likely to cause doubt to arise in Western authors.

Concordantly, as we have demonstrated, there was not a single direction until the scientific modernity, as the West or the Western Northern Hemisphere put it in a reductionist or discriminating way. HUTINGTON concludes in his work that what the West (Western Northern Hemisphere) considers universal, for the other geographies is Imperialism. The social does not change the logic or scientific methodology, but only its epistemological basis (GOLDEN), or mode of production that is oriented, for the specific branches of the respective knowledge (physics, calendars, etc.).

It is this magnitude of intellectual advance that makes some Western scientists doubt. The sociologist LATOUR, indicates that the excuse of the West (to behave even in an Imperialist way), is that the science of Nature, as it is, was previously discovered there. Which obviously falls down (it shows its ideological character if it is not changed), since the pre-Columbians also discovered Nature as it is. That author will finally conclude, that in those excluding terms 'we were never Modern' (western northern hemisphere). This is what he effectively demonstrates in principle, by avoiding discrimination. The history of Mayan Science, with respect to the West.

But even well-intentioned authors such as KROTZ (La Otredad cultural entre la utopía y la ciencia FCE 2002) or WALLERSTEIN (Conocimiento y saber el Mundo s XXI 2011), still continue to write respectively that Modernity was born in the Western city, or that there is only one world modern, which is that of Western capitalism. We also call this into question, but we are closer to what the anthropologist HARRIS once pointed out, that America represented a scientific experiment, and as such here we stick more to scientific veracity (if there are no major errors of this type in our work on the history of science).

It is already a few years ago that D. COE expressed that the true Science elaborated by the pre-Columbians was being discovered. We believe indeed that the Modern laws of temporal Nature are the crowning of this great process. As GALILEO referred (quoted by ARCINIEDAS, when he had in his hands a codex similar to that of the Mayas), 'there were before the Europeans, civilizations that studied Nature (stones, vegetables, stars').

Thus the Mayas belong to all America, but more deservedly to Latin America (Argentina), from where this scientific discovery (thesis) was not only produced, but demonstrated.

Bibliography

1. MERTON K. (1970): *Science in England in the 16th century.* Alliance p-20.
2. GOULD J. (2004): *Structure of the Theory of Evolution* Tusquet p-169.
3. PIAGET J. (1975): La Inteligencia Paidos; *Psicología de la Inteligencia* Psique p-64.
4. POPPER K. (1999). *La Lógica de la investigación CientificaTecnos* p-229.
5. VERNOUX-GODIN-BESSNNARD. (2019): *Mathematical plants Research* and *Science* p33.
6. COHEN I. (1983): *The Newtonian revolution* Alliance p-37.
7. GOODWIND B. (1998/· Tusquet *leopard spots* p-83.
8. WILCZEK F. (2016j.· *The World as a Work of Art* Drakonto p-177.
9. TREJO C. (1973): *Matemática Moderna* Eudeba p-105.
10. PIAGET J. (1972): *From the logic of the child to that of the adolescent* Paidos p-231.
11. AVENI A. (1980) (compiler): *Astronomía en América antigua* S. XXI.
12. MENZIES G. (2008): 1434 *Harper Collins* p-247.
13. PIAGET J. (1975): *Introducción a la Epistemología Genética Tomo II* Paidos p-51.
14. MONOD J. (1972): *Chance and Necessity Barral* p-114.
15. TAO T. (2019): *Universal Laws Monographic Issue Research and science* p-36.

General Bibliography

- ALSINA C. The sect of numbers Editée 2011.
- APLEYARD B Science and humanism Athenaeum 2003.
- AVENI A Astronomy in Ancient America Siglo XXI 1976; Observadores del cielo en México antiguo Fee 1961; La Astronomía Maya Mundo Científico Ciencia 1985.
- ARCINIEGAS G El revés de la Historia 1980.
- ALVAREZ C. Diccionario lingüístico Maya Unam 1980.
- ALBERDIJ. Selected works Quilmes University 2000.
- AUSTIN LTamoanchan-Tlalocan. Fee 1995.
- BRODAJ. (Compiler) Arqueoastronomia Mesoamericana Unam 1991.
- BROTHERSTON G Indigenous America in its literature Fee 1989.
- BELL F History of mathematics FCE 1996.
  BENITEZ-ROBLES De Newton y Newtonianos Quilmes 2006.
- CAMPI E Role sociologist in Argentina Science 1986; Integrated Psychology of Personality Development Editorial Académico Español 2016; The Scientific Modernity of the Pre-Columbian Maya EAE 2020; Discovery of Scientific Laws of Human Nature EA 2021.
- COE D The Maya Diana 1980; Decipherment of Maya Glyphs Fee 1995.
- COE M America Folio 1994; Mayas incognita and realities Diana 1990.
- COHEN E. Elements of sociology of knowledge Eudeba 1975.
- COHEN E The Newtonian Revolution Alliance 1984.
- CROMBIE A History of Science Alliance 1984.
- CID F History of Science Planeta 1979.
- COLERUS E Historia de la matemática Doncel 1972.
- CORRAL M. Historia de la Astronomía en México Ciencia 1986.
- DORADO R. La religión Maya Alianza 1986; Los Mayas una sociedad Oriental Complutense 1982.
- DI PAOLO History of AAA calendars 1989.
- DAVIES A: Enigmatic Mayan codices Andromeda 2006.
- ESCALONA R Cronología y astronomía Maya Mexica Fides 1940.
- ELIADE M Treatise history of religions Era 1975; Shamanism and techniques of ecstasy FCE 1986.
- ELENA A The Chimera of the Heavens XXI 1985.
- ESTRELLA History of Science Akal 1982.
- FULS A. The enigma of the Mayan calendar Science and research 2004.
- FERGUSON N Western Civilization and the Rest Debate 2021.
- FISCHETTI M Final climate argument Science and research 2019.
- FREUD S. Obras completes Biblioteca Nueva 1972.
- FERMI L What GALILEO Donsel said 1977.
- GRONDONA M Las conclusiones culturales del desarrollo económico Ariel 1993, Bajo el imperio de las ideas morales Sudamericana 1970.
- GOLDSTEIN A The dawn of science Fee 1984.
- GORTARI A La ciencia en Historia de México Fci 1963.
- GRIBBIN J History of Critical Science 2002,
- GOODWIN The spots of the leopardTusquet 1998.
- GOODY J Capitalism and modernity Critica 2004.
- GUILLEN M Equations that changed the world Debate 1999.
- GIRARD R Historia de las civilizaciones antiguas de América, Hyspamerica-Mexicanos 1970 ; El calendario Maya mexica. Mexico 1948.

- GOCKELW Decipherment of Maya script Diana 1995
- GUEVARA-PUIG The measures of the world Editée 2011
- GARZA M Los Mayas Inah 1968.
- HARDOY S. The pre-Columbian city Infinito 1999.
- HARRIS M General Anthropology Alianza 1982
- HOLTON G Studies in scientific thought. Alianza 1978.
- HAMOND N. Emergence of the Mayan culture Scientific American 1986
- HUTINGTON S Clash of civilizations Paidos 2002.
- HASSELKUS H WOOH Mexico 1993; Estructura glífica Maya Mexico 1998,
- HIDALGO I The Mayan computation Diana 1978.
- JUILLIEN Treatise on Efficiency Profile 1996,
- JOULET A The secret of numbers Science 1996.
- JONHSON P Birth of the Modern World Vergara 1999.
- KUHN TH The Essential Tension FCE 1980.
- KLINE M Mathematics for Humanities Fee 1972
- KOYRE A Studies of scientific thought S. XXI 1978; Galilean Studies S. XXI 1975; From the closed universe to the infinite Alliance 1978; From the world of lo approximately to the precise Critica 1948.
- KOESTLER THE Sleepwalkers Eudeba 1963.
- LIGNAIS F Grandes corrientes del pensamiento matemático Eudeba 1965.
- LINDBERG D Inicios de la ciencia Occidental Paidos 2002.
- MANCUSO S The future is vegetable Gutermberg 2017.
- MAROTTO F In the beginning was the Numero Bonalletra 2019.
- MATURANA H Maquinas y seres vivos S. XX 2004.
- MAUPONE L in Historia de la astronomía en Mexico Unan (compilation)1999; Reseña actividad astronómica en América antigua.
- MOYER G Gregorian Calendar Scientific American 1982.
- MERTON K Sociology of science Alliance 1997.
- MILLA VILLENA Ayni Anaru Waina 2004.
- MORLEY S. Maya hieroglyps Dover 1975 ; Antiqui Maya Sansoni-Firenze 1956; The Maya civilization Fee 1986.
- NEEDHAM J. Sociology of Science Alliance 1976.
- PORTILLA Time and reality in the Mayas Unam 1986.
- PIAGET-GARCIA History of Science S. XXI 1982.
- PRIGOGINE I. The New Alliance Alliance 1978
- POPPER K The open society and its enemies Paidos 1980.
- PICKOVER C From Arquinedes to Hawking Drakontos 2008 ; The Book of Mathematics Bookseller 2009.
- ROSSIJ America the great error of official history Galerna 1980.
- REY J.La revolución científica Icaria 1978.
- STRAUSS C. Structural Anthropology Eudeba 1978.
- SARTON G Six Wings Eudeba 1955.
- SEJOURNE M Nahuatl Thought in Calendars s. XXI 1999
- SEDEÑO E El rumor de las estrellas S. XXI 1973.
- STEWART I History of mathematics Drakontos 2008.
- SHAPIN R Scientific Revolution Paidos 2000
- SPRAJ C Venus rain and Corn Scientific 1998
- SCHELE-FREIDEL-PARKER Cosmos Maya Fee 1999
- SOUSTELLE J THE Mayas Fee 1966.
- SANTOS G The Mayas and the Ancient Empire Madrid 1981.

- SHARER The Mayan Civilization Fee 1998.
- TAO T.Universal Laws Research and Science 2015
- THOMPSON E Grandeza y decadencia de los Mayas Fee 1986 : Arqueología Maya Diana 1980;Comentarios al Códice Dresde FCE 1993; La religión Maya s. XXI 1988.
- THOMPKIN R The tree is born from hell Calvin 1992.
- TREJO C Matemática Moderna Eudeba 1973.
- VERNOUX-GODIN- BESNNNARD Plants that do mathematics Research and science 2019.
- WALLRSTEIN Impensar a las ciencias sociales S. XXI 2007.
- WHITEHEAD D Adventure of ideas Factory 1961.
- WEINBERG S Explaining the World Taurus 2016
- WILCZEK F The world as a work of art. Drakontos 2016.
- WULF A Invention of Nature. Taurus 2016.

AUTHOR'S DATA

Eduardo A. Campi: Child Psychologist (Universidad Autónoma de Entre Ríos), Professor of Psychology (Universidad Tecnológica Nacional, Colegio del Uruguay "Justo José de Urquiza") Concepción del Uruguay, Entre Ríos.

Study trips: Switzerland (invited by PIAGET Foundation), Mexico and Brazil (exhibitor on Mayan theme).

Awards and Recognition:

- 1987 - Kennedy School (Epistemology)
- 1998 - Arts and Sciences (Sociology)
- 1992 - FAGAP (Psychology)
- 2010 - Entre Ríos (Human Rights).

CONTRATAPA cover:

The Mayas in the II ac century, began to pass the premodern level and between the III-IV centuries of. ha were with their science, on the Modern level. In this book you will find a synthesis of the scientific Temporal Laws (Classical Calendar Science), which the Mayas wrote and enunciated about it.

Printed by Books on Demand GmbH, Norderstedt / Germany